The Power Of Virtualization: How It Affects Memory, Servers, and Storage

The Revolution In Creating Virtual Devices And How It Affects You

"An introduction to a new way of creating virtual computing resources that will change everything"

Dr. Jim Anderson

Published by:
Blue Elephant Consulting
Tampa, Florida

Copyright © 2016 by Dr. Jim Anderson

All rights reserved. No part of this book may be reproduced of transmitted in any form or by any means, electronic or mechanical, including photocopying, recording or by any information storage and retrieval system without written permission of the publisher, except for inclusion of brief quotations in a review.

Printed in the United States of America

Library of Congress Control Number: 2016916498

ISBN-13: 978-1539596462
ISBN-10: 153959646X

Recent Books By The Author

Product Management

- What Product Managers Need To Know About World-Class Product Development: How Product Managers Can Create Successful Products

- How Product Managers Can Learn To Understand Their Customers: Techniques For Product Managers To Better Understand What Their Customers Really Want

Public Speaking

- Tools Speakers Need In Order To Give The Perfect Speech: What tools to use to create your next speech so that your message will be remembered forever!

- How To Create A Speech That Will Be Remembered

CIO Skills

- Becoming A Powerful And Effective Leader: Tips And Techniques That IT Managers Can Use In Order To Develop Leadership Skills

- CIO Secrets For Growing Innovation: Tips And Techniques For CIOs To Use In Order To Make Innovation Happen In Their IT Department

IT Manager Skills

- Save Yourself, Save Your Job – How To Manage Your IT Career: Secrets That IT Managers Can Use In Order To Have A Successful Career

- Growing Your CIO Career: How CIOs Can Work With The Entire Company In Order To Be Successful

Negotiating

- Learn How To Signal In Your Next Negotiation: How To Develop The Skill Of Effective Signaling In A Negotiation In Order To Get The Best Possible Outcome

- Learn The Skill Of Exploring In A Negotiation: How To Develop The Skill Of Exploring What Is Possible In A Negotiation In Order To Reach The Best Possible Deal

Note: See a complete list of books by Dr. Jim Anderson at the back of this book.

Acknowledgements

Any book like this one is the result of years of real-world work experience. In my over 25 years of working for 7 different firms, I have met countless fantastic people and I've been mentored by some truly exceptional ones. Although I've probably forgotten some of the people who made me the person that I am today, here is my attempt to finally give them the recognition that they so truly deserve:

- Thomas P. Anderson
- Art Puett
- Bobbi Marshall
- Bob Boggs

Dr. Jim Anderson

This book is dedicated to my wife Lori. None of this would have been possible without her love and support.

Thanks for the best years of my life (so far)...!

Table Of Contents

1. INTRODUCTION 9
2. THE PROBLEM 10
3. VIRTUALIZING MEMORY 11
4. HOW VIRTUAL MEMORY WORKS 13
   - 4.1.1 Mapping Between Virtual Memory and Physical Memory 17
5. VIRTUALIZING SERVERS 19
   - 5.1 Why Bother With Virtualizing Servers? 21
   - 5.2 The Other Reason To Virtualize Servers 23
   - 5.3 The Role Of A Hypervisor In Server Virtualization 24
     - 5.3.1 What A Hypervisor Does 28
     - 5.3.2 Hypervisor Details 31
   - 5.4 Types Of Virtualization 34
   - 5.5 Server Virtualization In The Real World 35
6. VIRTUALIZING STORAGE 36
   - 6.1 How Computer Storage Started 36
   - 6.2 The Arrival Of NAS 38
   - 6.3 The Arrival Of SANs 39
   - 6.4 The Arrival Of Storage Virtualization 41
   - 6.5 Server Based Storage Virtualization 43
   - 6.6 Storage Network Based Storage Virtualization 46

6.6.1　In-Band Storage Network Based Storage Virtualization..47

6.6.2　Out-Of-Band Storage Network Based Storage Virtualization..48

6.6.3　Advantages And Disadvantages Of In-Band And Out-Of-Band Storage Network Based Storage Virtualization 49

6.7　Storage Controller Based Storage Virtualization 50

7　CONCLUSION 53

1 Introduction

Computers and networks have been fantastic inventions. With these tools, we have been able to do things that were never before possible. Immense amounts of data can now be processed, data can be stored anywhere and then either used or displayed at virtually any other location in the world. The arrival of smart phones and cloud computing means that we can now take sophisticated applications with us almost anywhere. However, we never seem to be satisfied with what we have – we always want more.

This desire for more, more, more is nothing new. Since the early days of computers we've always wanted to run multiple programs at the same time without interference, we've wanted to use multiple operating systems on a single server, we've wanted to mix and match storage solutions from different vendors, and we've wanted to be able to get more out of the networks that we build. The good news is that almost as soon as we create new things, we find ways to get additional performance out of them. In recent years, this has been accomplished through the use of virtualization.

Virtualization can take on many different forms: virtual memory, virtual servers, and virtualized storage. In each of these cases, we've been able to move beyond the underlying physical components that make up the system and through the use of clever virtualization techniques we've been able to get more out of the system. This is what we're going to explore in this book.

This book is divided into three sections. We are going to start out where all virtualization started: virtual memory. We'll then scale up and discuss how servers are virtualized. You can't have servers without storage, and so we'll then talk about how storage has been virtualized.

This is a rapidly evolving field. It our hope that by reading this book we can provide you with a firm foundation on which to understand what virtualization is and how it is being used to build networks that will be able to support today's applications. Using this foundation, we hope to be able to share with you the fundamental design approaches that are used to create a modern network and compare it to today's legacy networks.

2   The Problem

All great innovations come about because there is a problem that requires them. The field of virtualization came about because the world of Information Technology did a poor job of using what limited resources they had been given. Although virtualization had been around in various forms since the early days of computers (which really isn't all that long ago), virtualization had not taken off like it has today because there wasn't a big need for it.

Servers were cheap. Once a company's Information Technology (IT) team created a new application, additional storage needs were handled by purchasing additional servers and installing the servers in the corporate data center. Once the new hardware was "racked and stacked", it was plugged it in, with the new application loaded up, and the additional software functionality was off and running. The steep decline in the cost of server hardware beginning in the late 1990s and through the first quarter of the new century made it easier and easier to add more and more servers to corporate IT infrastructures.

This unparalleled, unchecked hardware growth began to slow once data center costs exponentially increased. Chief Information Officers (CIOs) and other senior managers took a step back and reviewed their data center costs. The assessment was not good.

Corporate data centers had expanded without constraints and were still requiring more space. All companies were going to have to make sizable investments in building new data centers. However, very few if any of the servers that were in the existing data center were being fully utilized. In fact, many of them were running at as little as 10% utilization. With the realization that this was an enormous waste of computing resources, it was also clear that something had to be done to provide needed resources, without excessive capitalization costs and with a sustainable, scalable growth plan. The answer is virtualization, but before this, a step back in time first.

3   Virtualizing Memory

Back in the 1950's – 1960's, when mainframes first started to show up, a problem quickly showed up. The programs that were being written to run on them were too large. When you took into account the software instructions that had to be executed, the variables that had to be stored, and all of the extra control information that the mainframe's operating system ate up, you very quickly ran out of memory. The IBM 650 computer had a memory size of 2,000 words (a word was 36 bits)[1].

Add to this the arrival of shared computing resources where more than one user wanted to make use of the same computer at the same time and you had a problem on your hands. Unlike today where it's a simple effort to jump into the car, run to the store, and pick up another Gigabyte of computer memory and then slap it into your system, back at the birth of mainframes memory was both expensive and limited – the computer could only handle so much of it. Clearly computer designers had a problem on their hands.

The way that they solved this problem was quite clever. What they did was to "virtualize" the computer's memory. The virtual memory technology that they invented lets a computer with a limited amount of physical memory to use its hard disk storage system along with its physical memory to make it look like the computer had a lot more memory to application software.

Initially, virtual memory referred to the idea of using the computer's hard disk to extend a computer's physical memory. The idea was that programs that were running on the computer would not have to care whether the memory was "real" memory [i.e., Random Access Memory (RAM)] or disk based. The operating system and hardware would figure that out. This solved the problem of trying to run programs that were larger than the computer's memory.

Later on, virtual memory was used as a means of memory protection. Every program uses a range of addresses called its address space. This solved the problem of allowing multiple users to use the same computer at the same time. The use of virtual memory prevents programs from interfering with each other. If a user's process tries to access an address that is not part of its address space, then an error occurs. The operating system will take over and will usually kill the process.

A computer that has been programmed to make use of virtual memory now has the additional task of managing its memory. It will start to inspect its RAM in order to determine what programs or data that has been loaded into its physical memory space has not been used recently. Once such areas have been identified, the CPU will then copy them from the RAM to the computer's hard drive. The RAM that this information had been using is now considered to be available for use by other applications.

The copying of RAM contents to the hard drive happens both automatically and very quickly. The result is that the end user and other applications don't even realize that it is happening. This means that the entire contents of the computer's RAM storage will appear to be available for each application. The lower cost of disk space over computer memory provides an additional economic incentive for this approach.

However, there are differences between the computer's hard drive and its RAM. The biggest difference is that the read / write speed of a hard drive is much shower that the read / write speed of the RAM. Additionally, hard drives have not been designed to support the accessing of small pieces of information. If there is too much interaction between the CPU and the hard drive in order to support the virtual memory function, then the end user and applications will observe a slowdown in performance. Let's take a look at how a virtual memory system can be successfully implemented.

4   How Virtual Memory Works

Virtual memory starts with a program that is too large to fully fit into the computer's physical memory. This program can be a stand-alone application, a user's online session with system configuration settings, or multiple users' applications. The first thing that the computer's operating system must do is to divide the single large application up into equal sized chunks of the original program (called "pages") that are generally between 4k – 64k in size.

The computer's physical memory (RAM) is then divided into equally-sized "pages". The memory addressed by a process is also divided into logical "pages" of the same size. When a process references a memory address, the memory manager fetches from disk the page that includes the referenced address, and places it in a vacant physical page in the RAM. This is shown in Figure 1. Pages that are not currently being used are stored on the hard disk. Subsequent references within that logical page are routed to the physical page. When the process references an address from another logical page, it too is fetched into a vacant physical page and becomes the target of subsequent similar references.

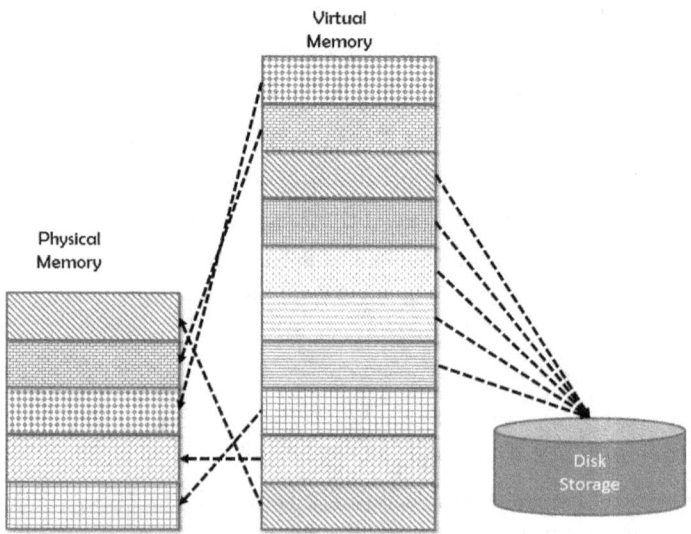

Figure 1: Virtual Memory To Physical Memory Mapping

If the computer does not have a free physical page when it comes time to swap in a virtual page, the memory manager can swap out a logical page into the hard disk's swap area, and copies (swaps in) the requested logical page into the now-vacant physical page. The page that was swapped out may belong to a different process. There are many strategies for choosing which page is to be swapped out. If a page is swapped out and then is once again referenced, it will be swapped back in from the swap area at the expense of another page that is currently in the computer's RAM.

This mapping allows the computer's main processor to view all of the memory that it is addressing as being part of one contiguous address space. In order for each program that is running on the computer to execute correctly, it is the responsibility of the operating system to manage the virtual address spaces and to assign real memory locations to virtual memory locations. The operating system has the ability to create a virtual memory space that is much larger than the amount of physical memory installed on the computer.

This management of the computer's virtual memory space is often performed by computer hardware called the Memory Management Unit (MMU). This hardware is responsible for automatically translating the virtual addresses that are being used by the executing applications to physical addresses.

The area of the hard disk that stores the RAM image is called a page file. It holds pages of RAM on the hard disk, and the operating system moves data back and forth between the page file and RAM. This process is shown in Figure 2.

The virtual memory system enables each process running on the computer to act as if it has the computer's entire memory space to itself since the addresses that it uses to reference memory are translated by the virtual memory mechanism into different addresses in physical memory. This allows different processes to use the same memory addresses on the same computer. The memory manager will translate references to the same memory address by two different processes into different physical addresses.

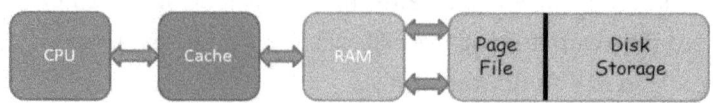

Figure 2: Memory Management Mapping

A process running on the computer may use an address space larger than the available physical memory. Each reference to an address from within the program will be translated into an existing physical address. The bound on the amount of memory that a process may actually address is the size of the swap area, which may be smaller than the addressable space.

An example of this would be a process that has an address space of 4GB yet actually only uses 2GB, and this can run on a machine with a page file size of 2GB.

The size of the virtual memory on a system must be smaller than the sum of the computer's physical RAM and the swap area. The reason for this limitation is because pages that are swapped in are not erased from the page file swap area, and so they take up two pages in the total available memory space.

The Windows operating system uses a swap area that is 1.5 times the size of the computer's RAM. This means that for a computer that has 4G of RAM, the swap area will be 6G and the total size of the virtual memory will be 10G.

4.1.1 Mapping Between Virtual Memory and Physical Memory

The key to a successful virtual memory system is to create a process by which the computer can use virtual addresses to execute a program while the operating system handles the details of mapping between the application's virtual address space and the computer's physical address space.

Every program in a computer's memory, along with its data, is called a "process". Within the computer, each process is given its own address space. An address space is a sequence of valid memory addresses that can be used by the process to store both code and data. An address space is not fixed. While a process is executing, it can request additional memory from the operating system. Once it is done with the additional memory, it can then be given back to the operating system.

As a process executes, it generates addresses in one of three different ways: a load instruction, a store instruction, or by fetching the next instruction to be executed. These addresses that are being generated by the process are considered to be "virtual addresses". In order for actual data or instructions to be fetched or stored, this virtual address will then have to be mapped to a real physical address.

What this means is that computers that is using virtual memory will be performing the virtual to physical memory mapping function all the time. In order to make this process as efficient as possible, it is general handled by special purpose hardware.

There is an added benefit to using virtual memory when more than one process is executing on a computer. The virtual memory system provides a process with memory protection via address translation. This means that it's worth the extra hardware that is required in order to get memory protection.

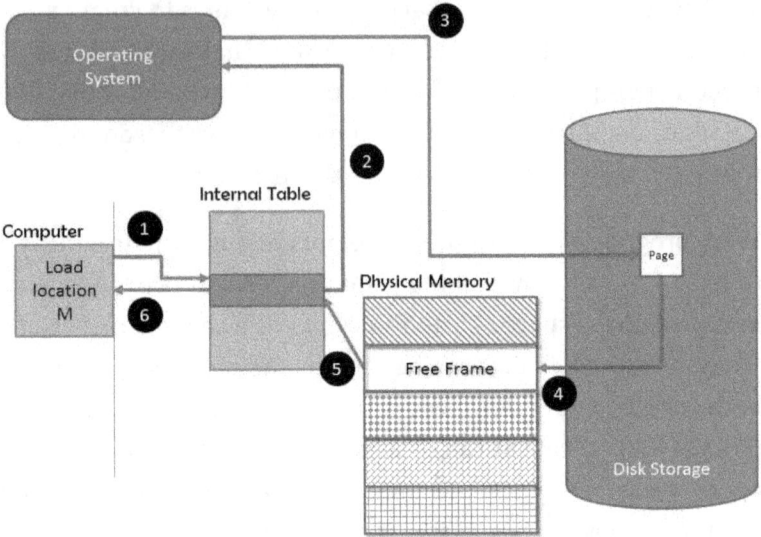

Figure 3: How Virtual Memory Is Mapped To Physical Memory

4.1.2 The Memory Mapping Process

Figure 3 shows how virtual memory is mapped to physical memory. It is a 6-step process. The steps are as follows:

- **Step 1:** Check an internal virtual memory address look-up table for this process. Determine if the reference was a valid request or it was an attempt at an invalid memory access.

- **Step 2**: If the virtual memory address reference was invalid, terminate the process. If it was valid, but the page has not yet been brought in, page in the requested page.

- **Step 3**: Find a free frame in the computer's RAM.

- **Step 4**: Schedule a disk operation to read the desired page into the newly allocated RAM frame.

- **Step 5**: When the disk read is complete, modify the internal table that is kept with the process and the page table in order to indicate that the page is now in memory.

- **Step 6**: Restart the process instruction that was interrupted by the address trap. The process can now access the page as though it had always been in memory.

The extra effort that is required to implement a virtual memory system is generally considered to be worth it because of the advantages that it delivers. A virtual memory system will provide applications with increased security due to memory isolation, frees applications from having to manage a shared memory space, and will allow applications to use more memory than might be physically available, using the technique of virtual memory paging.

5 Virtualizing Servers

With the success that they had had with virtualizing computer memory, scientists were ready to tackle their next computer challenge: how to allow multiple people to use a computer at the same time. Original computers were very expensive and they could only do one thing at a time. Users would queue up and wait to have their program loaded, run, and the results printed out. All of this happened in a sequential fashion. Clearly, as computers became more and more vital to the way that businesses and governments operated, this inefficiency had to be resolved.

The solution to this problem that was created was once again quite novel: the entire computer was virtualized. The initial foray into computer virtualization took the form of the creation of a "time-sharing" mainframe system and culminated in the development of the CP-40 operating system. Each user was provided with a virtual machine (VM), which enabled multiple users to access the same mainframe computer simultaneously (See Figure 1). A software hypervisor was created to manage memory sharing in the mainframe. Every user was fooled into believing that they were the only user on the mainframe computer at any given time. Meanwhile the mainframe raced around frantically doing small jobs for each user in a sequential fashion.

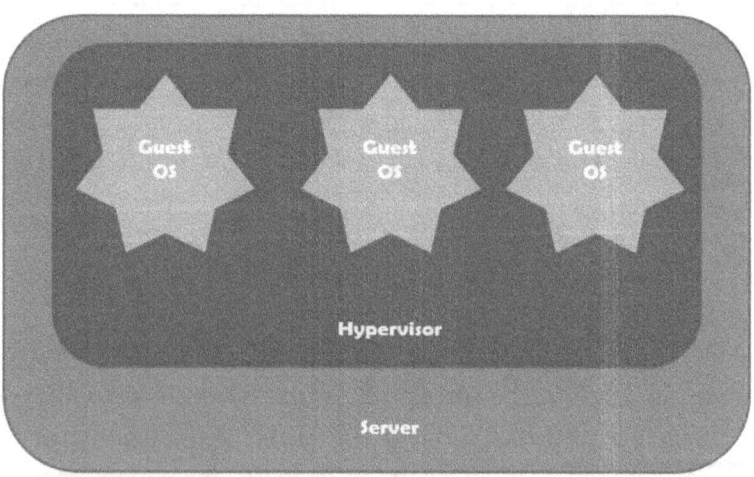

Figure 4: Server Virtualization

5.1 Why Bother With Virtualizing Servers?

With the arrival of personal computers, the need to use server virtualization techniques faded away. These new computers supported only a single user and there was no longer a need to support the overhead that virtualization required. That one user made use of all of the personal computer's resources and their needs were met.

Time moved on and personal computers become more powerful. They started to be used by themselves, no human user was involved. These enterprise class servers were being placed in racks which were then placed in data centers. Soon hundreds of servers were being used to run the wide variety of applications that a modern business uses every day.

Many of today's servers live in data centers. These data centers are responsible for making sure that the servers have power 7x24x365. In fact, making sure that all of the servers in the data center always have power is so important that most data centers end up creating two identical but separate power systems just in case something happens to one of them. As you might guess, this can lead to a great deal of waste.

At the request of The New York Times, the consulting firm McKinsey & Company analyzed energy use by data centers and found that, on average, they were using only 6 percent to 12 percent of the electricity powering their servers to perform computations [4].

How did things get to be so bad? It all goes back to how things started out. Back in the early 1990s software systems crashed if they were asked to do too many things, or even if they were turned on and off. In response, computer technicians seldom ran more than one application on each server and kept the machines on around the clock, no matter how sporadically that application might be called upon. Can you see where I'm going with this one?

A crash or a slowdown could end a career. Data center operators live in fear of losing their jobs on a daily basis. This means that they'll make sure that every computer is kept up and running no matter how much (or how little) it is being used. In technical terms, the portion of a computer's brainpower being used on computations is called "utilization."

McKinsey & Company has been monitoring the issue since at least 2008. The server utilization figures have remained stubbornly low: the current findings of 6 percent to 12 percent are only slightly better than those in 2008.

Just because you can virtualize a server, does not necessarily mean that you should virtualize a server. There has to be a reason to go ahead and introduce the added complexity into the operation and management of the servers that a business or a government is using. It turns out that that there is a very good reason for doing this: it can save a lot of money.

What firms have discovered is that they are paying a lot of money to keep servers that they are not using very efficiently up and running all the time. Shutting these many servers down and running a smaller number of servers that have been virtualized to allow multiple applications to run on them at the same time makes very good business sense.

5.2   The Other Reason To Virtualize Servers

Cost savings and reducing the number of servers that a business needs in order to run its business are very good reasons to implement server virtualization. However, there is more to this story than just that. There is no question that virtualizing file, print, and web servers makes good business sense. These applications can't fully utilize a single server. However, there are additional benefits to implementing virtual servers.

In today's IT departments, they must find better ways to integrate, provision, deploy, and manage IT systems — at a faster pace — without further straining already tight budgets. Greater optimization and efficiency is needed in how software and solutions that power data centers are deployed and managed.

What this means is that having the ability to create a new virtual machine with only the press of a button can result in incredible savings in both time and expense. This task is so simple to do, in some cases the creation of new virtual servers can be turned over to the actual end users to take care of.

An additional driver for server virtualization is that firms want to cut down on the rogue activities of some of their departments. These departments are known for going off on their own, ordering server equipment as needed, and then turning them over to IT to manage once they've arrived. In order to both streamline and optimize the way that a company's IT infrastructure looks, servers need to be virtualized and then handed out on an as needed basis.

One of the most important points to keep in mind is that virtualization is not a goal by itself. A much better way to think about virtualization is to view it as being a means to the strategic goal of enabling services-based IT in the enterprise. Thought of this way, we can all view virtualization as being a journey — not a destination.

5.3 The Role Of A Hypervisor In Server Virtualization

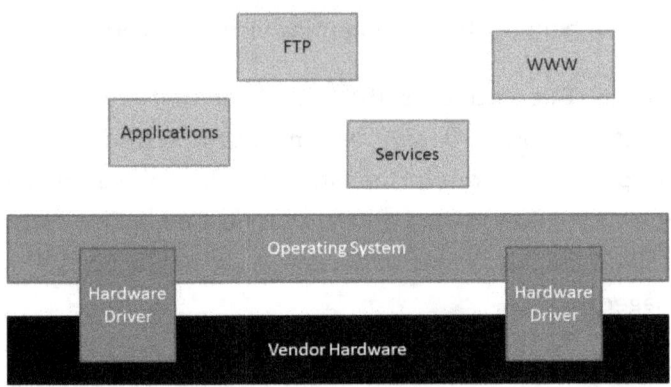

Figure 5: Traditional Model Of A Non-Virtualized Server

Figure 5 shows the components of a non-virtualized server. In this model, vendor provided hardware is used as the basis for the entire server system. On top of this is added an operating system. The operating system interacts with the underlying hardware via a set of hardware drivers that are customized to the specific vendor's hardware that is being used. On top of the operating system, a collection of various applications and services can be run independent of the vendor hardware that is being used.

In this model, only one operating system can be used on a hardware platform at a given point in time. Additionally, the functionality that the operating system is able to provide to the upper layer applications and services is limited by the functionality that is provided to the operating system by the hardware drivers. Note that changing an operating system or the underlying hardware in this configuration is very difficult to do.

A side effect of using this model is that a company often ends up with a collection of underutilized servers ("server sprawl") because most software vendors state that their applications should run on stand alone servers in order to avoid software conflict issues with other applications.

In Figure 6, the architecture of a virtualized server is shown. This type of virtualization is known as "platform virtualization". A new layer of software has been added in this figure: a virtualization layer (also known as a "hypervisor") that sits between the vendor hardware and the various operating systems that will be running on this virtualized server. A large number of vendors currently offer virtualization layer software (VMware, Microsoft, Citrix, Oracle, etc.) along with popular open source solutions (i.e. Xen, Hyper-V, KVM, etc.).

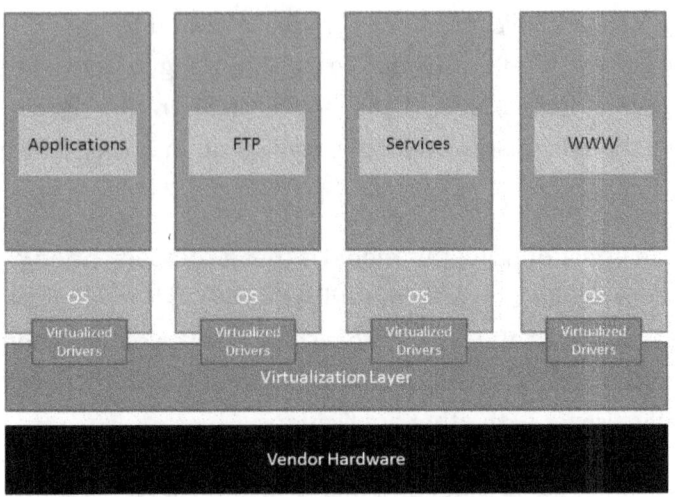

Figure 6: Architecture Of A Virtualized Server

The virtualization layer will have the needed drivers embedded in it which will control the underlying hardware. The virtualization layer then presents generic server hardware to the operating system: generic graphics cards, generic network interfaces, generic storage systems, etc.

On a virtualized server there will be several differences from the non-virtualized server. The first of these changes is that multiple operating systems can be run on the same physical hardware at the same time. Each one of these operating systems will be using a piece of the available total computing resources that are available. Each of the virtual environments ("virtual machines") is both encapsulated and isolated. It cannot impact any of the other virtual machines while it is running. No virtual machine will be allowed to monopolize the physical server or any of its resources.

The operating systems can no longer "see" the actual hardware that they are running on – the virtualization layer has make all of the hardware appear to be generic hardware. This means that each of the virtual machines is now portable. Since each of the virtual machines is defined by a set of files, they can be easily moved from one physical server to another as long as the new machine is running the same virtualization layer.

When virtual machines have been backed up, they can be restored to any hardware. This means that in the event of the hardware failure of a server in a data center, a new server from a different vendor can be installed to replace the failed server. Once the virtualization layer of software is installed on the new vendor hardware, the backups of the virtual machines that were running on the old vendor's hardware can now be restored and they will continue to operate with no changes – the underlying hardware no longer matters.

The ability to restart a virtual machine is a fundamental feature of most virtualization layers. In the event of the failure of a virtual machine, the virtualization layer can automatically restart the virtual machine. If the entire server fails and the virtualization layer has been networked to other virtualization layers in the data center and shared storage is being used, the virtual machines can be restarted on other servers.

Virtualization layers are able to measure the utilization of the servers that they are running on. This allows additional virtual machines to be continuously added to the server until some predefined level of utilization is reached. It is recommended that a buffer be left in this utilization level (say 60%) in order to allow for spikes in individual virtual machines processing loads. Note that this is much better than the estimated 4%-7% utilization of most non-virtualized servers that are found in today's date centers [16]. In this way companies can ensure that each of the servers in their data center is being fully utilized.

Implementing virtualized servers in a data center allows servers to be consolidated. This occurs when the applications that used to run on multiple servers are now combined and start to run on a single server. Server consolidation ratios of 7x (seven physical severs replaced by one physical sever running a virtualization layer) or 10x are not uncommon.

5.3.1   What A Hypervisor Does

At the heart of any sever virtualization is the virtualization layer of software, or as it's more commonly referred to as the hypervisor. A hypervisor is a piece of software that creates the layer of virtualization that makes virtualization on a server possible. The hypervisor contains the Virtual Machine Manager (VMM) which has the responsibility of managing each of the active virtual machines that are currently running on the server.

There are two different types of hypervisors. Both types are shown in Figure 7. The first, Type 1 ("Bare Metal"), is loaded directly onto the server hardware (i.e. "bare metal"). Examples include Hyper-V, VMware's ESX / ESXi, and Xen. This software interacts directly with the server hardware with no layer of software between it and the server hardware. Type 1 hypervisors provide true partition isolation, high reliability for virtual machines, and a high degree of security.

The other type of hypervisor, Type 2 ("OS Hosted"), is loaded into the operating system which is already running on the server hardware. Examples include VMWare Workstation, Microsoft Virtual Server, and VMWare Fusion. This type of hypervisor then interacts with the underlying operating system which then interacts with the server hardware.

Type 2 hypervisors may initially appear to be easier to install than Type 1 hypervisors. Simply put, they are just an application that is run on a standard server – no special software installation is required. However, Type 2 hypervisors do not perform as well as Type 1 hypervisors simply because the operating system is between the hypervisor and the physical hardware. Type 2 hypervisors provide low cost virtualization, do not require the use of any additional drivers, and are easy to both install and use.

This means that there is additional software overhead when you use a Type 2 hypervisor which results in the restriction that it is not possible to fit as many virtual machines onto a server that is using a Type 2 hypervisor. Ultimately, the consolidation ratio that can be achieved with a Type 2 hypervisor is smaller than that can be achieved with a Type 1 hypervisor.

Type 1 hypervisors are generally used in data centers. They are used to maximize the number of physical severs that can be consolidated onto a single virtual server. Type 2 hypervisors are more often used on laptops and desktops when there is a need to run multiple operating systems or create two separate development environments for testing or development purposes.

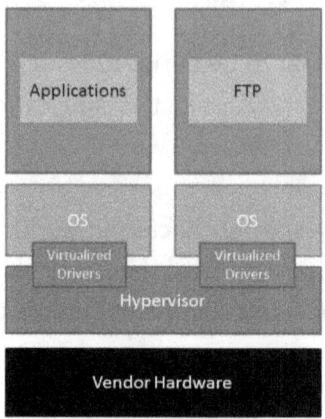
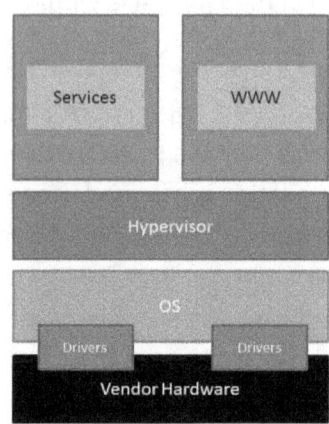

Figure 7: Types Of Hypervisors

The hypervisor views the server as providing four main types of resources. These resources are the CPU, memory (RAM), storage (hard disk), and a network connection. The hypervisor then uses these three types of resources to provide the various virtual machines with the functionality that they need. This includes infrastructure services such as memory management, storage management, and network data exchange management.

Server consolidation will result in reduced costs for the company that is using the servers. These costs will include reduced energy costs, reduced software license costs, reduced data center costs, and reduced maintenance contracts. At the same time disaster recovery will be simplified and improved as will server manageability.

All hypervisors use a "control domain" to manage the hypervisor and the virtual machines that are running on it. This is a special purpose virtual machine that has the ability to manage the hypervisor. In some hypervisors such as Xen, this is also where the virtual drivers execute and device models are stored. The management console that the system administrator uses to configure the virtual server is connected to the control domain [15].

The hypervisor management system is the key to accessing the functionality that the virtualization system provides. The basic functions of creating, halting, and terminating virtual machines are provided. Additionally, the ability to create live snapshots, checkpoints, and migrate virtual machines to other servers is also provided. While a virtual machine is running it can be migrated. This migration can be between host servers or pools of host servers including those that don't have shared storage.

5.3.2   Hypervisor Details

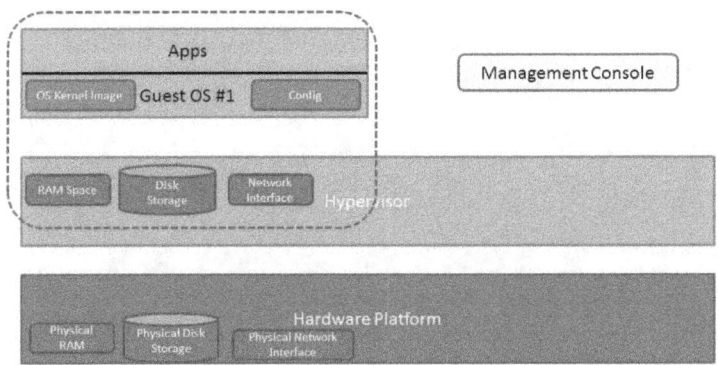

Figure 8: Typical Hypervisor Resources

A hypervisor is a special type of application that serves to abstract the underlying physical server hardware from the guest operating systems that are executing in each of the virtual machines that are running on the virtual server. Figure 8 shows the typical resources that a hypervisor has. A hypervisor allows each application running on the server to see the resources provided by its virtual machine instead of the actual physical server resources.

One of the fundamental tasks of a hypervisor is to permit a guest operating system to boot on the virtual server. In order to permit this happen, the hypervisor needs to provide the OS kernel image (Windows, Linux, Solaris, etc.) with a configuration file that tells it what IP addresses can be used, how much RAM is available, etc. Additionally, access to a hard disk for file storage and a network card for external communication also needs to be provided. The virtual disk and virtual network interface will be mapped by the hypervisor into the server's physical disk and network interface. The one other component that is required is a management console that permits both the hypervisor and the virtual machines to be controlled.

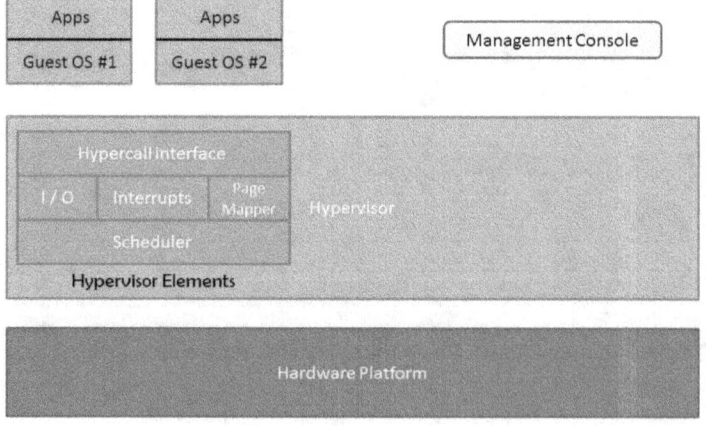

Figure 9: Detailed View Of A Hypervisor

The architecture of a hypervisor is designed to allow guest operating systems to be run concurrently with the hypervisor. Figure 9 shows the hypervisor elements that are required in order to make this happen. In typical applications, calls will be made to the operating system in order to perform functions on the server's physical resources (RAM, hard disk, network card, etc.) The hypervisor provides the hypercall layer in order to allow guest operating systems to make requests of the server's physical resources. Likewise, when an app attempts to use I/O devices attached to the server, the hypervisor will virtualize the request and map it to the physical device.

Interrupts pose a special challenge for the hypervisor. Interrupts for virtual devices have to be routed to the guest operating system. Additionally, during the execution of an app in a virtual machine, traps or exceptions can occur. When this happens, the impact needs to only affect the virtual machine where the trap / exception occurred and not the hypervisor or any other virtual machines.

Within the hypervisor, the page mapper component is responsible for ensuring that the server's instruction fetching hardware always points to the correct page for the virtual machine (or the hypervisor) that happens to be executing at that time. Who is currently executing is the responsibility of the hypervisor's scheduler component which is responsible for transferring execution control between the hypervisor and the various virtual machines that are executing.

Resource pools can be created that contain both flexible storage and networking resources. Within the virtual environment, events can be tracked and progress notifications can be sent. Upgrades to the virtual environment can be made and patches can be applied without having to take the virtual environment down. Real-time performance monitoring is provided and alerting capabilities are provided.

5.4 Types Of Virtualization

There are four different ways to virtualize a server [17]. Each one of these approaches uses a different configuration of the three virtualization components: applications, operating systems, and hypervisors.

- **Full Virtualization**: When full virtualization is used, the hypervisor is responsible for completely simulating the underlying vendor hardware. This permits unmodified copies of operating systems (e.g. Windows, Linux, etc.) to execute on the virtualized server within their own virtual machines.

- **Hardware-Assisted Full Virtualization**: As virtualization has become both more popular and more critical to the efficient operation of a data center, CPU manufactures have responded by adding instructions to their products that support virtualization. When these virtualization-enabled CPUs are used to power a server, the hypervisor can leverage their features to allow guest operating systems to operate in complete isolation. One feature of these CPUs is the introduction of the "ring" concept in which refers to levels of security privileges that are permitted to the application that is currently executing. Applications operate at a ring 3 level, rings 1 & 2 are used to execute device drivers, and ring 0 is used to execute the hypervisor. AMD and Intel have also created a ring -1 that permits the hypervisor to run computations directly instead of going through the operating system. The result of this is that there is an increase in the efficiency of the processing.

- **Paravirtualization**: In a virtual server that is using paravirtualization, the guest operating systems have each been modified to inform them that they are operating in a virtual environment. Paravirtualization permits the relocating execution of critical tasks from the virtual domain to the host domain. The result of this is that they will spend less time performing operations that are more difficult in a virtual environment compared to a non-virtualized environment.

- **OS Virtualization**: When OS virtualization is being used on a server, a hypervisor is not used at all. Instead, the virtualization capability is built into the host OS. The host OS performs all the functions of a fully virtualized hypervisor. The biggest limitation of this approach is that all the VMs must run the same OS. Each VM remains independent from all the others, but it is not possible to mix and match operating systems among them.

5.5   Server Virtualization In The Real World

The original problem that firms ran into with the servers that they were using to run their enterprise applications was that it simply took too many servers to do the job. When enterprise applications were created, they were designed to run on dedicated, purpose-built servers for maximum performance and stability. This was a good idea, but it lead to what was called "server sprawl" where an expensive data center could quickly get filled up with servers that were only being partially used. This is the problem that server virtualization was created to solve.

The invention of server virtualization technology now allows enterprises to consolidate both applications and operating systems (often jointly referred to as "workload" when talking about virtualization) that are running on multiple unrelated servers. Now these can all be run on a single server that is executing multiple virtual environments (VEs).

The use of virtualization allows an enterprise to run multiple applications on one single physical server. Each application executes in a separate VE. This allows each application to believe that it is running on a purpose-built server. Allowing each application to continue to execute on top of its specific operating system eliminates a potential source of interoperability issues that could occur between applications that were required to share an operating system.

6 Virtualizing Storage

Just as virtualization has had a dramatic impact on the world of servers, the world of storage needed a similar type of revolution. The amount of data that is being generated on a daily basis that has to be stored has grown every year.

EMC and the International Data Corporation together estimated that more than 1.8 trillion gigabytes of digital information were created globally in 2011 [4]. Just generating and storing this data is not enough. The stored data has to be made available to servers so that software applications can process it. This is where the revolution in storage technology is occurring.

6.1 How Computer Storage Started

The revolution in computer storage first got started way back in the era of mainframes. System designers had a problem: the magnetic disks that the mainframes were using were not all that reliable and failures were all too common. They needed to find a solution to this problem.

Their solution was to take large numbers of magnetic disks and pool them together. They then arranged them in a way that would provide fault tolerance to protect against individual disk failures. This system became known as RAID (Redundant Array of Independent Disks). Applications could use a common pool of cache memory in order to work with a logical image of a data block rather than having to work with the actual data block as it was spun around on a disk platter. The end result of this design was that it improved performance by masking the seek and rotation delays of a mechanical disk. An added benefit was that it allowed mainframes to use lower cost disks.

About this time, desktop computers started to show up in the early 1980's. Computer storage was fairly simple to implement for these systems. A hard drive would be added to the computer system, a cable would be run between the hard drive and the rest of the computer and voila! The stored information could be accessed by whichever program happened to be running. This type of solution was called Direct Attached Storage (DAS).

Of course, the need for storage of the information that we've generated just keeps growing and this was true even back in the early days of computers. Very quickly users started to exceed the storage limitations of their DAS solutions. The next step was for standard protocols (the language used to communicate between devices) to emerge. Examples of this included the small computer system interface (SCSI). One of the strengths of this standardized way of connecting a computer to a storage system is that it made it easy to connect different vendor storage solutions to the systems.

Where we actually stored our data continued to change also. The use of SCSI interfaces extended connectivity beyond traditional magnetic storage to devices such as CD-ROMs, tape drives and autoloaders, and JBOD (just a bunch of disks). Storage solutions continued to evolve and soon fault-tolerant designs were available which provided greater reliability. However, the connectivity of these storage systems was still confined to single servers or workstations, limiting the utilization of the media. If you wanted to access a specific data set, you needed to sit down and use a specific computer.

6.2   The Arrival Of NAS

The limitation of only being able to access a data set from a single computer very quickly became a major problem that had to be solved. The world of computer networking was exploding at this time: expensive proprietary solutions were all giving way to a fast, simple solution called Ethernet. The ability to network multiple servers together led companies such as Novell and Sun Mircrosystems to start to wonder if the new high speed networking could be used to solve the limitations of the current storage systems.

Their research into this problem resulted in the creation of what is called Network Attached Storage (NAS). What NAS allowed computers to do was to treat the storage that it was using as though it was physically attached to the server even if it was in a different room or perhaps a different city. This functionality was a huge breakthrough: now data could be stored at a central location, but it could then be accessed from many diverse locations. The ultimate benefit of this is that a NAS system would allow users to more closely collaborate. Very quickly other vendors started to include NAS functionality with their products.

6.3 The Arrival Of SANs

For a period of time, NAS solutions solved the problem of allowing multiple users to access remotely stored data by many different users. However, problems still remained. One of the biggest storage problems that firms were dealing with was that they now had a large number of separate NAS systems. Whenever a new data set had to be stored or accessed, a new NAS was created. This got out of hand very quickly.

What companies needed was a way to pull together all of their separate NAS solutions into a single storage solution. This is why Storage Area Networks (SANs) were created. SANs helped organizations consolidate their storage assets in order to improve capacity utilization by sharing their storage resources effectively. The result of this was that they were able to simplify storage management by using common software tools, and it enabled the replication of critical information over long distances to provide greater levels of protection against data corruption and disaster events.

So what is a SAN? A SAN is a storage architecture that has been designed to connect detached computer storage devices, such as disk arrays, tape libraries, and optical jukeboxes, to servers in a way that the devices appear as local resources. SANs can then deliver storage to servers at a block level, and provide feature mapping and security capabilities in order to ensure that only one server can access the allocated storage at any particular time. The protocol, or language, used to communicate between storage devices and servers in a SAN is SCSI (small computer system interface)

The use of SANs helped companies to consolidate their storage assets in order to improve storage capacity utilization by sharing their storage resources more effectively. This resulted in simplified storage management by allowing the use of common software tools, and enabled replication of critical storage information over long distances in order to provide greater levels of protection against data corruption and disaster events. Many larger organizations, such as banks and telecommunication providers, were among the first to see the value in this and implement SANs.

SANs were not the perfect storage solution. As you might imagine, although the technology was powerful, it was how it was implemented that often caused a new set of storage-related problems for enterprises. One of these problems was that "islands" of SANs were being created. The reason for this is that there was very little interoperability between SAN products that were created by different vendors and storage devices.

Even after a SAN was implemented by an enterprise, people were discovering that storage utilization levels were still fairly low. The reason for this was that traditional pre-SAN storage allocation methodologies were still being used. When this was coupled with the fact that companies could not replicate or move their storage between storage solutions provided by different storage vendors this ended up limiting the storage options that were available to companies.

6.4 The Arrival Of Storage Virtualization

Storage virtualization is the process by which the physical storage of data is abstracted in order to create logical storage. The physical storage systems, disk drives, are first aggregated into "storage pools". These storage pools are then used to create the logical storage entities which will then be presented to the applications that want to store and use data and which can be managed from a single console.

There are numerous reasons for virtualizing storage. These include being able to offer applications more flexible storage options along with simplified management of the stored data. As is the case with virtualized servers, once storage has been virtualized, better capacity utilization of the storage systems combined with better performance can be provided to the applications that use the virtual storage system.

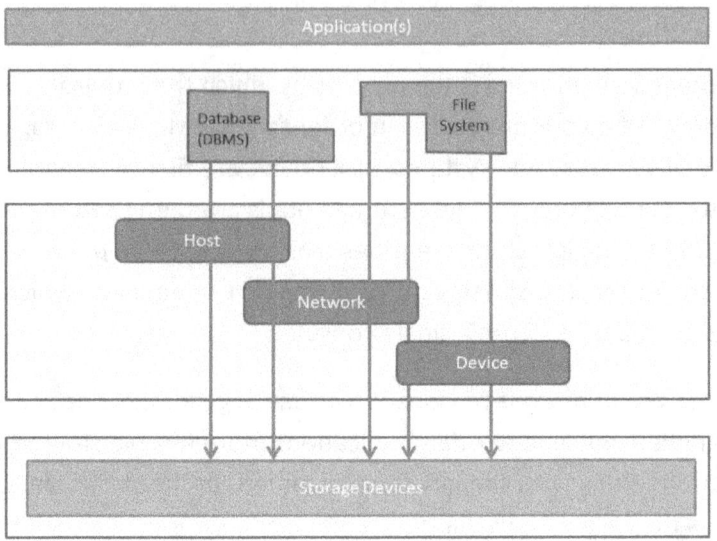

Figure 10: A Shared Storage Model

A shared storage model has been created in order to show how applications actually store data onto physical storage systems [18]. This model is shown in Figure 10. This model consists of four separate layers. At the bottom there are the physical servers that the data will ultimately be stored on. Sitting on top of that layer is the Block Aggregation layer which contains the host, network, and device interfaces. On top of that is the File / Record layer and at the very top is the application layer.

When a storage system is virtualized, the virtualization can occur at any layer. The layer at which the virtualization occurs then presents a virtualized view of the storage system to the layer above it. There are three different types of storage virtualizations that can be implemented: server, storage network, and storage controller.

6.5 Server Based Storage Virtualization

Storage virtualization got its start in the operating systems that run on servers. At a physical level, data is being stored on a hard disk in fixed sized blocks that could be read or written. However, keeping track of what data was located where on a given hard disk quickly became a challenge.

The operating system implemented a rudimentary virtualization technique where a file name was associated with a group of storage blocks. Now by using a file name, an application could easily find the data that it wanted to reference. As the number of files grew, eventually the available storage on a given hard disk was exhausted and additional hard disks (or "volumes") had to be connected to the computer in order to provide additional storage.

As with all things related to storage, soon the number of volumes that server had to access in order to perform its tasks quickly grew to become unmanageable. Operating systems were then modified to support the idea of a Logical Volume Manager (LVM). An LVM is simply a group of volumes that have been grouped together into storage pools. This is shown in Figure 11.

Since the application no longer had to worry about which volume it was retrieving its data from, this allowed the concept of "data striping" to be introduced to storage systems. When data striping is being used, data that is being written out to a storage system is split between multiple volumes. Every disk has a built in delay when it is asked to retrieve data while it performs a seek to find the start of the requested data on its storage platter. When the data is spread across multiple disks, this delay operation can be performed in parallel and then all of the data can be delivered in the time that it takes to read one segment.

Server based storage provides three key benefits:

1. **Stand Alone**: No additional hardware has to be purchased in order to implement server based storage virtualization. Any storage system that the server is able to access can be virtualized using this approach.

2. **Cheap**: The software required to implement server-based storage virtualization is already part of the operating system that is being used on the server. This means that there is no additional software that has to be purchased or licensed.

3. **Flexible**: Since server-based storage virtualization is part of the operating system, it is very easy to configure and is extremely flexible.

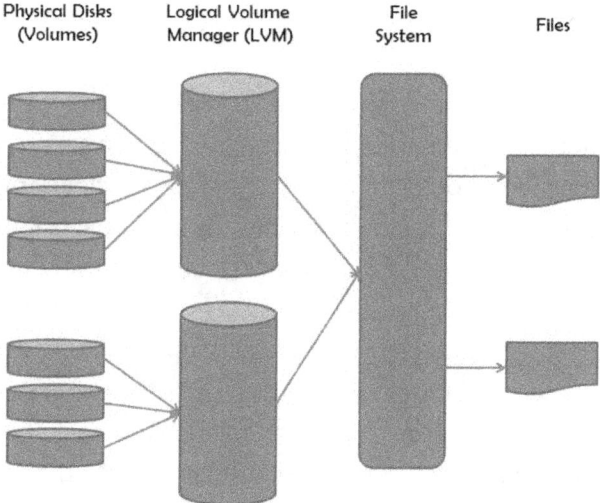

Figure 11: How Logical Volume Managers Are Used To Store Information

Using server based storage virtualization does have its own set of limitations. These limitations may only affect certain configurations and usage circumstances. The limitations are as follows:

1. **Data Migration**: When data has to be migrated or replicated across storage systems in order to provide redundancy it can very quickly become a challenge to keep track of the level of data protection that is being provided.

2. **CPU Slowing**: Virtualizing the storage system of a server takes a great deal of CPU processing. As data flows both to and from the server subsystem, it needs to be mirrored, striped, and parity bits need to be calculated. What this means is that this will take computing resources away from the applications that are being processed on the server resulting in a loss of

performance.

3. **Single Server**: Server based storage virtualization is, by necessity, limited to working with a single server. It will maximize both the resilience and the efficiency of that single server, it won't provide any advantages for other servers that need to access the same data set.

4. **Unique File Systems**: The way that file systems are implemented on a storage system is dependent on the vendor that provided the storage. This means that for each storage system there is the possibility that a unique virtualization solution will have to be implemented and then maintained.

6.6   Storage Network Based Storage Virtualization

The arrival of Storage Area Networks (SANs) allowed the job of managing the storage and retrieval of data from a storage system to be moved off of the server and onto special purpose-built hardware. Now the servers communicated with the SAN system anytime they needed to store or retrieve data. It wasn't long before vendors realized that the SAN would be a good place to implement storage virtualization.

There are two types of storage network based storage virtualization: in-band and out-of-band. Both types are show in Figure 12.

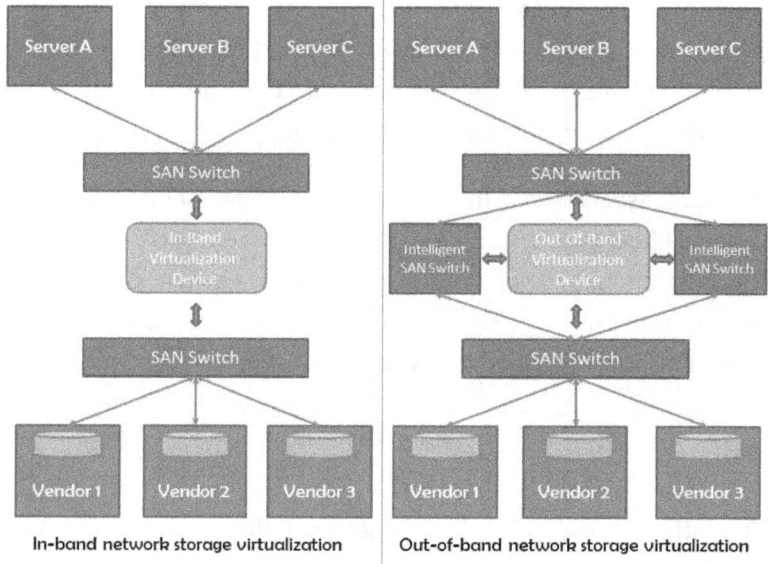

Figure 12: Types of Storage Network Based Storage Virtualization

Both types of storage network based storage virtualization abstract the view of the physical storage systems between the server and the SAN. The only difference is if they perform this function in-band or out-of-band.

6.6.1   In-Band Storage Network Based Storage Virtualization

When in-band storage network based storage virtualization is being used, the virtualization function becomes part of the communication path that exists between the server and the SAN as shown in Figure 12. This functionality can either be built into a SAN switch or provided by an additional piece of hardware.

In an in-band solution, the server will never see the physical storage system. It will only see the virtualization device / functionality. It is the responsibility of the in-band virtualization function to accept storage requests from the server, analyze it, perform a look-up on its storage mapping tables in order to discover where the requested data is to be stored or read from, and then perform the storage operation.

A unique aspect of the in-band solution is that all of the storage data will flow through the in-band component. This means that it is possible to provide data on storage data usage, cache storage data, manage any replication services that are being used, and manage data migration operations.

6.6.2 Out-Of-Band Storage Network Based Storage Virtualization

The difference between out-of-band solutions and in-band solutions is that in an out-of-band solution, the virtualization component does not lie in the path of the storage data. Instead, as shown in Figure 12, the virtualization component is connected to virtualization enabled SAN switches that perform all of the required look-ups.

Once again, the servers have no direct contact with the storage system. Instead, the servers interface with the SAN switches which are then responsible for interfacing with the physical storage system. In this type of a solution, the out-of-band virtualization component maintains a meta-data map of the stored data and the storage system. This map is then used to tell the SAN servers where to go in order to save or retrieve the data that the servers have requested.

In an out-of-band solution, the storage data never passes though the out-of-band virtualization component. This reduces the amount of delay that virtualization introduces into the system; however, it also means that the storage data cannot be cached. Because of this, the out-of-band solution cannot boost the performance of the storage system.

6.6.3 Advantages And Disadvantages Of In-Band And Out-Of-Band Storage Network Based Storage Virtualization

As with all solutions, there are advantages and disadvantages that the in-band and out-of-band solutions provide to the applications that use them. There are three primary advantages of using this type of storage virtualization solution:

1. **Pooling**: Using either type of solution allows storage solutions from multiple vendors to be pooled into a seamless storage pool that is accessible by all servers.

2. **Replication**: Stored data can now be replicated over storage solutions that come from multiple different vendors.

3. **Management**: The ability to manage the complete storage solution can be provided by a single management console.

Likewise, there are drawbacks to this type of solution. The three most significant drawbacks are as follows:

1. **Complexity**: These are not easy storage solutions to set up. The physical storage elements have to be mapped to their virtual counterparts. This is done by creating a mapping table which then becomes a potential single point of failure. The custom nature of these mapping tables means that vendors are able to lock in their customers once the table has been created. Moving your data from a virtualized storage solution to a

different storage solution can be difficult or impossible.

2. **Maintenance**: The virtualization device is generally a high powered server that will require as much maintenance and software updates as any other server. The number of servers involved will grow as more are added to permit clustering solutions that will provide the needed backup and reliability that this part of the storage solution will demand. As the amount of data being stored grows, the servers may be hard pressed to scale with the solution.

3. **Latency**: In both of these types of solutions, the server's request to the storage solution no longer travels directly to the physical storage system. Instead, there is now a collection of SAN and virtualization devices in between the server and the storage system. This means that delay will be introduced to every interaction between the server and the storage system. Couple this with the limited amount of CPU cycles and RAM that the SAN and virtualization devices have and you have the potential for delay to be introduced into the system.

6.7 Storage Controller Based Storage Virtualization

In a storage controller based storage virtualization solution the vendor provided storage array provides the virtualization services. This type of solution allows heterogeneous vendor storage arrays to be connected to the vendor's storage controller. Once this is done, all of the storage arrays are managed by the servers as though they were internal disk drives.

This type of solution can be implemented with a great deal less complexity that a Storage Network Based Storage Virtualization solution. No additional layer of management software is required. The real value of this type of solution is that the storage controller is able to virtualize the storage and this allows each of the storage arrays to be used by the server as though they were a part of the storage solution no matter where they may be physically located.

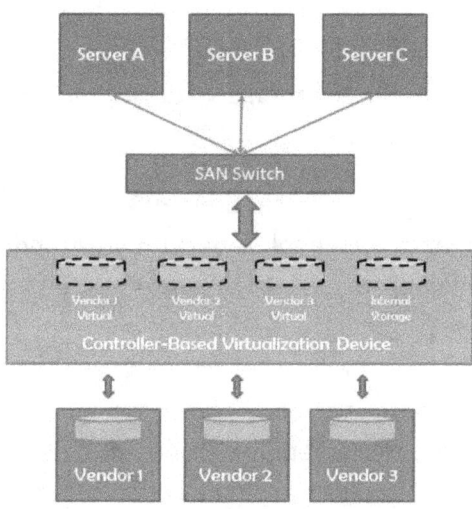

Figure 13: Controller Based Storage Virtualization

A storage controller based virtualization solution allows all of the functionality provided by the storage controller to be extended to all of the storage components provided by all of the vendors. The real value of this is that data can be replicated between storage pools with no disruptions to availability of the data. Additionally, replication can occur between storage systems provided by different vendors.

Storage resources can be allocated to specific applications. This means that specific cache, ports, and disk pools can be made available to a specific application. This can allow an application to achieve a given level of service or implement a security policy.

In an enterprise storage solution, the use of a Storage Controller Based Storage Virtualization solution provides more disaster recovery options. Clustering capability functionality between different controllers has been created in order to permit the real-time recovery from the failure of a complete storage array failure.

Storage Controller Based Storage Virtualization provides five very clear benefits to users:

1. **Standards**: The way that servers are connected to external storage pools is performed via industry standard protocols. This means that users can avoid any type of proprietary vendor lock-in.

2. **Simplicity**: This solution allows the amount of SAN hardware that is required to be reduced. This means that the complexity of the solution is also reduced.

3. **Cheaper**: This solution builds on top of the existing SAN infrastructure and therefore is often cheaper than other storage virtualization solutions. Additionally, the tools required to manage the storage solution (availability, replication, management, etc.) can be consolidated.

4. **Legacy Support**: The most important storage system functions including provisioning, migration, replication, and partitioning can now be performed on legacy

storage equipment from other vendors.

5. **Data Protection**: The ability to replicate stored data between storage pools that are located on different classes of storage equipment or even equipment from different storage vendors means that overall data protection costs can be lowered and data protection solutions expanded.

7 Conclusion

The revolution in virtualization technologies has come about because of real needs that have arisen as memory, servers, and storage systems have all increased in both size and complexity. As computers have become a more and more important part of our lives, the applications that run on them have expanded in size and now need more processing power and more data in order to operate correctly. This has caused these applications to exceed the capabilities of the systems that they initially executed on.

Clever engineers have always been a step ahead of the software developers. As programs become larger and single computers were shared among multiple users, virtual memory techniques were designed and deployed. As the number of servers that were filing up data centers increased even as the utilization of these severs remained low, server virtualization was introduced. The popularity of this solution lead to computer chips being redesigned so that they could support server virtualization with new low level microprocessor commands. Finally, as our ever increasing need for more and more data expanded, multiple forms of storage virtualization have been introduced into our networks.

All of our computing needs have not yet been met. Data centers are still expensive to build and maintain, the amount of data that our programs need continues to expand. The solutions that we have suit us well for our needs to today, but tomorrow will demand new solutions. The only way that we'll be ready to create these new solutions is if we take the time to fully understand how we've come this far. Virtualization is only getting started, we can only imagine what the future holds for both us and our computers.

Photo Credits:

Cover - http://klarititemplateshop.com/
https://www.flickr.com/photos/ivanwalsh/

Other Books By The Author

Product Management

- What Product Managers Need To Know About World-Class Product Development: How Product Managers Can Create Successful Products

- How Product Managers Can Learn To Understand Their Customers: Techniques For Product Managers To Better Understand What Their Customers Really Want

- Product Management Secrets: Techniques For Product Managers To Boost Product Sales And Increase Customer Satisfaction

- Product Development Lessons For Product Managers: How Product Managers Can Create Successful Products

- Customer Lessons For Product Managers: Techniques For Product Managers To Better Understand What Their Customers Really Want

- Product Failure Lessons For Product Managers: Examples Of Products That Have Failed For Product

Managers To Learn From

- Communication Skills For Product Managers: The Communication Skills That Product Managers Need To Know How To Use In Order To Have A Successful Product

- How To Have A Successful Product Manager Career: The Things That You Need To Be Doing TODAY In Order To Have A Successful Product Manager Career

- Product Manager Product Success: How to keep your product on track and make it become a success

Public Speaking

- Tools Speakers Need In Order To Give The Perfect Speech: What tools to use to create your next speech so that your message will be remembered forever!

- How To Create A Speech That Will Be Remembered

- Secrets To Organizing A Speech For Maximum Impact: How to put together a speech that will capture and hold your audience's attention

- How To Become A Better Speaker By Changing How You Speak: Change techniques that will transform a speech into a memorable event

- How To Give A Great Presentation: Presentation techniques that will transform a speech into a memorable event

- How To Rehearse In Order To Give The Perfect Speech: How to effectively rehearse your next speech to that your message be remembered forever!

- Secrets To Creating The Perfect Speech: How to create a speech that will make your message be remembered forever!

- Secrets To Organizing The Perfect Speech: How to organize the best speech of your life!

- Secrets To Planning The Perfect Speech: How to plan to give the best speech of your life

- How To Show What You Mean During A Presentation: How to use visual techniques to transform a speech into a memorable event

CIO Skills

- Becoming A Powerful And Effective Leader: Tips And Techniques That IT Managers Can Use In Order To Develop Leadership Skills

- CIO Secrets For Growing Innovation: Tips And Techniques For CIOs To Use In Order To Make Innovation Happen In Their IT Department

- Your Success As A CIO Depends On How Well You Communicate: Tips And Techniques For CIOs To Use In Order To Become Better Communicators

- What CIOs Need To Know About Working With Partners: Techniques For CIOs To Use In Order To Be Able To Successfully Work With Partners

- Critical CIO Management Skills: Decision Making Skills That Every CIO Needs To Have In Order To Be Able To Make The Right Choices

- How CIOs Can Make Innovation Happen: Tips And Techniques For CIOs To Use In Order To Make Innovation Happen In Their IT Department

- CIO Communication Skills Secrets: Tips And Techniques For CIOs To Use In Order To Become Better Communicators

- Managing Your CIO Career: Steps That CIOs Have To Take In Order To Have A Long And Successful Career

- CIO Business Skills: How CIOs can work effectively with the rest of the company!

IT Manager Skills

- Save Yourself, Save Your Job – How To Manage Your IT Career: Secrets That IT Managers Can Use In Order To Have A Successful Career

- Growing Your CIO Career: How CIOs Can Work With The Entire Company In Order To Be Successful

- How IT Managers Can Make Innovation Happen: Tips And Techniques For IT Managers To Use In Order To Make Innovation Happen In Their Teams

- Staffing Skills IT Managers Must Have: Tips And Techniques That IT Managers Can Use In Order To Correctly Staff Their Teams

- Secrets Of Effective Leadership For IT Managers: Tips And Techniques That IT Managers Can Use In Order To Develop Leadership Skills

- IT Manager Career Secrets: Tips And Techniques That IT Managers Can Use In Order To Have A Successful Career

- IT Manager Budgeting Skills: How IT Managers Can Request, Manage, Use, And Track Their Funding

- Secrets Of Managing Budgets: What IT Managers Need To Know In Order To Understand How Their Company Uses Money

Negotiating

- Learn How To Signal In Your Next Negotiation: How To Develop The Skill Of Effective Signaling In A Negotiation In Order To Get The Best Possible Outcome

- Learn The Skill Of Exploring In A Negotiation: How To Develop The Skill Of Exploring What Is Possible In A Negotiation In Order To Reach The Best Possible Deal

- Learn How To Argue In Your Next Negotiation: How To Develop The Skill Of Effective Arguing In A Negotiation In Order To Get The Best Possible Outcome|

- How To Open Your Next Negotiation: How To Start A Negotiation In Order To Get The Best Possible Outcome

- Preparing For Your Next Negotiation: What You Need To Do BEFORE A Negotiation Starts In Order To Get The Best Possible Deal

- Learn How To Package Trades In Your Next Negotiation

- All Good Things Come To An End: How To Close A Negotiation - How To Develop The Skill Of Closing In Order To Get The Best Possible Outcome From A Negotiation

Miscellaneous

- The Internet-Enabled Successful School District Superintendent: How To Use The Internet To Boost Parental Involvement In Your Schools

- Power Distribution Unit (PDU) Secrets: What Everyone Who Works In A Data Center Needs To Know!

- Making The Jump: How To Land Your Dream Job When You Get Out Of College!

- How To Use The Internet To Create Successful Students And Involved Parents

- How Software Defined Networking (SDN) Is Going To Change Your World Forever: The Revolution In Network Design And How It Affects You

The Revolution In Creating Virtual Devices And How It Affects You

This book has been written with one goal in mind – to show you how the revolution in virtualization is going to affect you! This changes everything, let's makes sure that you are ready.

Let's Prepare You For The Arrival Of Virtualization!

What You'll Find Inside:

- **HOW VIRTUAL MEMORY WORKS**

- **WHY BOTHER WITH VIRTUALIZING SERVERS?**

- **SERVER VIRTUALIZATION IN THE REAL WORLD**

- **STORAGE NETWORK BASED STORAGE VIRTUALIZATION**

Dr. Jim Anderson brings his 25 years of real-world experience to this book. He's spent over 25 years working in the Telecommunications industry and teaching at Universities. His insights will show you how virtualization will impact your life and your network.

www.ingramcontent.com/pod-product-compliance
Lightning Source LLC
Chambersburg PA
CBHW070357190526
45169CB00003B/1037